High-Velocity V8s

y sincere thanks to the following people
r their time, information, images and
thusiasm for this book:

erry Justus and the Globecam team,
elbourne, Australia

mothy Madsen, Sydney, Australia

mie Whincup, Gold Coast, Australia

eve Owen, Brisbane, Australia

artin Whitaker, V8 Supercars, Brisbane,
ustralia

ole Hitchcock, V8 Supercars, Brisbane,
ustralia

ewart Mailer, V8 Supercars, Brisbane,
ustralia

3 Supercar teams: Jamie Whincup, Steve
wen, James Courtney, Steven Johnson

Dear Reader

During one of my school visits, some students suggested that I write a book about V8 Supercars. I didn't know very much about this motorsport so I contacted a friend, Gerry Justus, who manages a company called Globecam in Melbourne. Gerry's company designs in-car camera systems for V8 Supercars. With Gerry's help, I attended two V8 Supercar Championship races, where I thoroughly enjoyed interviewing V8 Supercar drivers and their teams. Everyone was very helpful. That's me pictured with Jamie Whincup in the picture opposite.

V8 SUPERCARS DRIVERS MUST BE VERY FIT AND HEALTHY TO DRIVE ABOUT 1000 KILOMETRES IN EACH CHAMPIONSHIP RACE.

I also liked the fact that V8 Supercars drivers help promote passenger and driver safety to school students around Australia.

Join me on a V8 Supercars tour!

Sharon Parsons

Contents

High-Velocity V8s

1 **V8 Supercars Championship**

The world's best drivers race for glory. page 4

2 **What Is a V8 Supercar?**

Super cars with extraordinary speed ... page 6

3 **V8 Supercars Drivers**

... but they're just cars, without drivers! page 8

4 **Jamie Whincup, from Go-Karts to V8s**

How Jamie became a V8 Supercar driver. page 10

5 **Jamie Whincup, Top V8 Supercar Driver**

Meet Jamie Whincup, a top driver. page 12

6 **Steve Owen, Elite V8 Supercar Driver**

Meet Steve Owen, one of the best! page 14

7 **Steve Owen Gets Ready to Race**

Detailed preparation is essential! page 16

8 Inside Steve Owen's V8 Supercar

Take a look inside a supercar. page 18

9 Vision from V8 Supercars

What do the drivers see? page 20

10 The In-Car Camera Systems

A high-tech system for fantastic images. page 22

11 V8 Supercars Trucks

Incredible trucks, incredible races. page 24

12 A V8 Supercars Event – New Zealand

Live from the track. page 26

13 V8 Supercars Safety Messages

It's not just about speed – safety is vital. page 28

EXPOSITION FEATURE

More Motorsports Should Switch to Bioethanol Fuel

TEXT TYPE Exposition

page 30

Index and Glossary page 32

1 V8 Supercars Championship

Every year in February, the V8 Supercars Championship begins at Abu Dhabi, in the United Arab Emirates. Hundreds of millions of people around the world attend or tune in to watch some of the world's best drivers in a series of V8 Supercars battles, which run from February to December. It's a gruelling travel schedule for everyone involved: there are 15 Supercars events around Australia, and two international events (Hamilton, New Zealand and Abu Dhabi, United Arab Emirates).

V8 SUPERCARS™

racing at night in Abu Dhabi

Racing for the corner!

Racing around the corner!

2 What Is a V8 Supercar?

A V8 Supercar starts out as a standard vehicle but most of the parts are specially modified to transform it into a high-performance Supercar.

Lightweight for Speed

To enable the Supercar to reach speeds of up to 300 kilometres per hour, all non-essential car parts are stripped out to make the car lighter.

Strong for Safety

A very strong roll-bar cage is fitted to protect the driver and strengthen the car's chassis. The car's extra strength also makes it faster.

Following the earthquakes in Christchurch, New Zealand, in 2011, V8 Supercars drivers showed their support for people affected by the disaster with signs on their cars.

Spoon is a V8 Supercars mechanic who enjoys his job.

AN EXPENSIVE SPORT

V8 Supercars racing is an expensive sport. Each car can cost about $500 000 to build so a V8 Supercar team relies on company sponsorships to pay for the additional costs, such as the driver's salary, running costs and a team of specialist support people.

In 2011, V8 Supercar driver Jamie Whincup won his fourth V8 Supercars Clipsal 500 race, in Adelaide Australia.

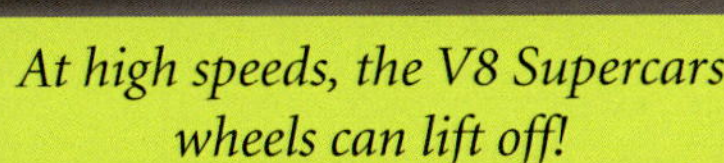

At high speeds, the V8 Supercars wheels can lift off!

Specifications for **Supercars**

Each V8 Supercar ...

- has a specially designed five-litre V8 engine.
- is among the heaviest racing vehicle in the world, at 1355 kilograms!
- can reach speeds in excess of 294 kilometres an hour!
- can accelerate to 100 kilometres an hour in 3.8 seconds!
- can hold 120 litres of bioethanol fuel (pages 30 and 31 have more information on fuel).
- includes a car-to-pit telemetry system (page 21 has more information on telemetry).

V8 ENGINE

In simple terms, a V8 engine has eight cylinders. Two rows of four cylinders are arranged in a "V" shape.

a computer-generated image of a V8 engine

3 V8 Supercars Drivers

Each year, only 28 drivers are selected to compete in the top tier of the of V8 Supercars Championship. The drivers are highly skilled and can drive at speeds of up to 300 kilometres an hour. They are extremely fit and are able to race for about two hours while maintaining focus and alertness in an effort to be first past the chequered flag.

V8 Supercars Drivers, 2011

Steve Owen

Jamie Whincup

A Featured Driver, Pages 10–13

Jamie Whincup

A Featured Driver, Pages 14–19

Steve Owen

A Driver and His Son

V8 Supercar driver Steve Johnson with his son

Aim for the Chequered Flag!

4 Jamie Whincup, from Go-Karts to V8s

Jamie Whincup is an elite V8 Supercar driver who is passionate about racing. He has been dedicated to his sport from the age of seven when he first raced go-karts.

"I always did my homework first and then couldn't wait to work on my go-kart with my dad," says Jamie.

Jamie Whincup

Jamie Helps Out at the Track

Jamie as a young boy, with his father (right) and uncle

> "I MAKE THE MOST OF RACE OPPORTUNITIES."
> JAMIE WHINCUP

Jamie believes that much of his racing success is due to his parents. He explains that his parents did not have much money for the expensive sport of go-kart racing, but they sacrificed a lot to support Jamie's passion and talent. Since the age of 17, Jamie has raced in the tough sport of competitive car racing, managing to advance to the top of his ultimate motorsport, V8 Supercars.

Jamie races his go-kart.

Jamie holds a well-earned winner's trophy – his father and uncle proudly pose with him!

Jamie's parents continue to support him at V8 Supercars events.

5 Jamie Whincup, Top V8 Supercar Driver

Jamie's team tent, where supplies for drivers and mechanics are kept during races

Focused on Race Strategy

Concentrating on the Race Ahead

Racing in the Rain

> WHEN I AM COMPETING IN A RACE AND I FEEL CONFIDENT THAT MY CAR HAS NO ISSUES, IT IS EASY TO STAY FOCUSED – I AM REALLY PASSIONATE ABOUT MY SPORT.
>
> JAMIE WHINCUP

A Super Career!

In 2002, Jamie Whincup began his V8 Supercars racing career. Follow his exceptional racing achievements from 63rd place to 1st place between 2002 and 2011.

At the time of writing this book, Jamie Whincup was leading in the 2011 V8 Supercars Championship.

Racing to Win

Year	Place
2002	63rd
2003	27th
2004	50th
2005	16th
2006	10th
2007	2nd
2008	1st
2009	1st
2010	2nd
2011	1st

Jamie makes history by consecutively winning the Adelaide, Australia, race three times.

Biography Snapshot

Born: 6 February 1983 in Melbourne, Australia
Lives: Gold Coast, Australia
Nickname: J-Dub
Favourite Circuit: Bathurst
Best Race Preparation: I love to go to the Murray River and think about my racing strategy!
Greatest Personal Influence: My dad has always given me good advice.
Interests: water skiing and trail bike riding
Favourite Food: chocolate cake

Jamie Whincup's Team

Jamie's race team works tirelessly to ensure that his car is in the optimum condition for performance and safety – before, during and after each race day.

Tony Monks is Jamie's head mechanic.

After every race, Jamie's team works like a complex human machine as they clean, maintain and test his car.

Jamie takes a break after the race.

6 Steve Owen, Elite V8 Supercar Driver

Steve Owen is a successful V8 Supercar driver who lives in Brisbane, Australia. Ever since he started racing go-karts at the age of ten, he has wanted to race V8 Supercars. When Steve turned 25 years old, he achieved V8 Supercars status.

Steve Owen admires a motorbike before he starts his V8 Supercars Championship race in Adelaide, Australia, in 2011.

Steve Owen's Racing Career

Year	Age	Racing
1984–1994	10–20 years old	Go-Karts
1995–1998	21–24 years old	V8 Driving Championships
1999–2010	25–35 years old	**V8 Supercars Development**
2011–present	**36 years old–present**	**V8 Supercars Championship**

Steve Owen racing in Hamilton, New Zealand

In the pits, Steve Owen quickly takes over the driving from Jamie Whincup while the car is refuelled and tyres are changed.

Endurance Race

In 2010, Steve Owen was asked to join Jamie Whincup as his co-driver for the Australian Phillip Island V8 Supercars endurance race (500 kilometres). Although they did not win, they did post the fastest qualifying time to take pole position on the starting grid. Great effort!

Sports

V8 Supercars Development

The second tier of V8 Supercars racing is called "V8 Supercars Development". Each year, the seven events in the series provide drivers with the training and experience that will enable them to challenge for a spot at the elite or top level. At this level, drivers race in 15 events.

The Author Interviews Steve

How many kilometres do you race each year?

I race more than 16 000 kilometres every year.

Why do you love V8 Supercars racing?

It's no different to any other sport that you're passionate about. I always try to get the best out of the car every time.

Do you race in all weather conditions?

The only time we don't race is when the track is flooded. For wet weather racing, special grooved tyres are fitted to the cars.

7 Steve Owen Gets Ready to Race

In the final minutes before a race, Steve Owen is focused on his gear and the race ahead. Follow his final pre-race moments before the championship race on a very wet track in Hamilton, New Zealand (April 2011).

Inside Steve's Tent

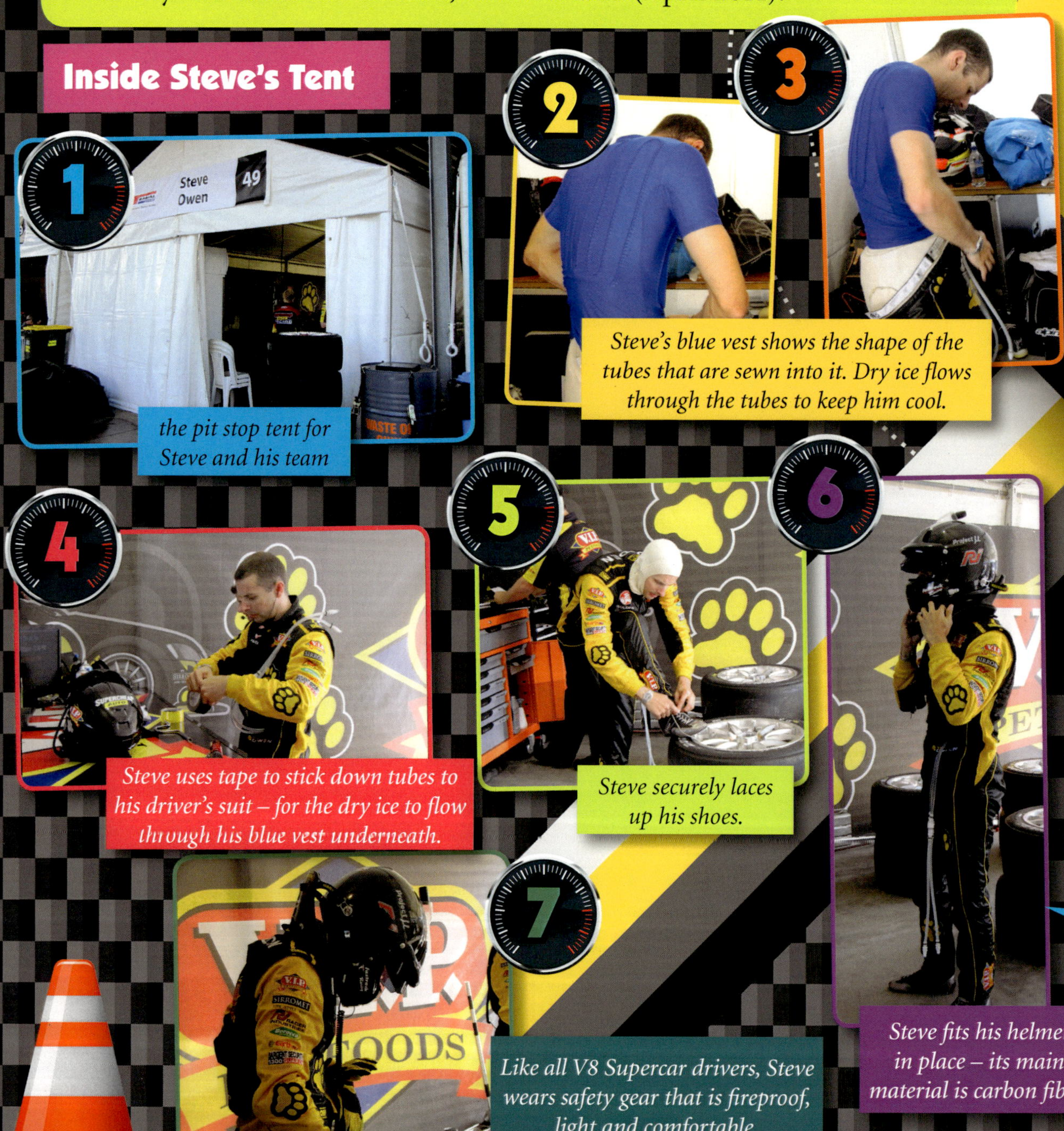

1 the pit stop tent for Steve and his team

2, 3 Steve's blue vest shows the shape of the tubes that are sewn into it. Dry ice flows through the tubes to keep him cool.

4 Steve uses tape to stick down tubes to his driver's suit – for the dry ice to flow through his blue vest underneath.

5 Steve securely laces up his shoes.

6 Steve fits his helmet in place – its main material is carbon fibre

7 Like all V8 Supercar drivers, Steve wears safety gear that is fireproof, light and comfortable.

Final Preparations in the Pit Lane

CARBON FIBRE

Carbon fibre is a light, strong and fireproof material that has many uses for sport-related clothing and equipment.

8 Inside Steve Owen's V8 Supercar

Extreme Conditions

Driving V8 Supercars is harder than it looks – it is extremely hot and noisy and there are no luxuries. There is no air conditioning so … it is an extra 20 to 25 degrees Celsius hotter inside the car than the temperature outside. If it is a hot 38 degrees Celsius outside, V8 Supercars drivers swelter in temperatures above 58 degrees Celsius!

Dry Ice Cool Box

Inside every V8 Supercar is the essential cool box filled with dry ice, which is located to the left of the driver. A tube connects the cool box to the network of tubes that are sewn into the driver's suit. The dry ice flows through the tubes to keep the driver cool.

inside Steve Owen's V8 Supercar

COOL BOX CAMERA

A miniature camera is fixed to the cool box. It can be remotely controlled by Globecam staff based at the Supercar event in a different location.

The cool box camera can be remotely moved along the miniature track by Globecam staff at the racetrack.

Cameras Transmit for Television

Six miniature cameras have been fitted inside Steve's car and they are powered by a small, lightweight battery. Live images are transmitted from the in-car camera system to Globecam's on-site computers and the outside broadcasting unit via fibre-optic cables, a car antenna and track antennas. The director at the outside broadcasting unit chooses the best vision to broadcast live on television.

A miniature camera transmits vision of Steve's head.

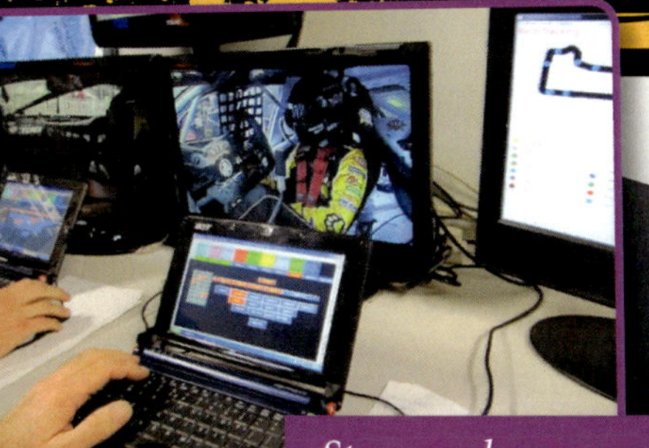
Steve can be seen on the television screens.

the vision switching system

fibre-optic cables at the track

antenna on top of Steve's V8 Supercar

a rear-facing camera inside a carbon-fibre housing structure

a close-up of Steve Owen as viewers see him on television

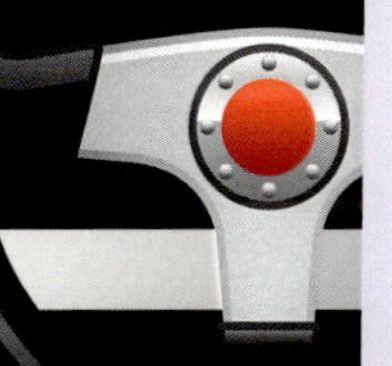

Health

Dehydration

To prevent dehydration, Steve has a four-litre bottle of water, which he sips on through a tube. People can only absorb about 750 millilitres of water each hour so he aims not to drink more than that amount while racing.

9 Vision from V8 Supercars

Globecam is a company made up of a talented team of Australian engineers and electronic specialists who design and manufacture in-car camera systems for V8 Supercars. Their work enables millions of spectators and viewers to see the action up close from inside and outside nine V8 Supercars, and around the track, too.

This screen shot shows the remote users interface for the V8 Supercars camera control system.

the Globecam team at a V8 Supercars event

the Abu Dhabi V8 Supercars track with Globecam camera and antenna checkli

Globecam team member Clint studies the layout of the camera system and antennas at the Abu Dhabi V8 Supercars track.

TELEMETRY FEATURE

Meet Tim

Tim owns a motorsport electronics company. His telemetry expertise and technology enable the V8 Supercars television team to access all kinds of accurate real-time information about the drivers and their cars.

At each V8 Supercars event, Tim often works alongside the Globecam team – they are also great mates!

from left: Scott, Gerry and Tim

TELEMETRY

Telemetry information is used widely in motor racing. It utilises wireless communications to collect, measure and report information from a remote location.

Tim listens to the director who works at the on-site V8 Supercars television outside broadcasting studio. Tim coordinates a lot of information on race day to help the director and his team.

TELEVISED WORLDWIDE

The V8 Supercars television team works with Globecam and other media crews to broadcast each event live to 110 countries and about 850 million people.

the V8 Supercars outside broadcast television studio

10 The In-Car Camera Systems

Gerry Justus and his team at Globecam are innovators and inventors because they design and build integrated remote-control camera systems for V8 Supercars. Their electronic systems in nine V8 Supercars and their fibre-optic cabling around the racetrack provide action-packed vision to the on-site V8 Supercars outside broadcast television studio. For most of the year, the Globecam team works in Melbourne, Australia, but they also travel to each V8 Supercars Championship event, too.

Gerry fits a camera and a vision-switching system to a V8 Supercar.

Scott uses the microscope to build the tiny electronic components.

Clint works on the design for part of a new integrated system.

Electronics Design

Sam is an electronics specialist at Globecam. One of his many tasks is to design, build and test printed circuit boards. The printed circuit boards enable the remote-control camera system to transmit signals from the cameras inside the V8 Supercars to the Globecam computers at the racetrack. At each V8 Supercars event, Sam is on hand to ensure that all the signals are transmitted and received because there cannot be a break in transmission at a live broadcasting event.

Sam designs complex printed circuit board systems on the computer.

Design Challenges

One of Globecam's biggest challenges is to design and build reliable, light and strong in-car camera systems. A car's performance must not be affected and a driver's vision must not be obscured during each race.

A camera, inside its housing structure, is bolted to the car's frame.

a tiny camera inside a headlight cavity

The cameras must be protected and securely bolted in places, such as under light covers, on the car's strong chassis and on the roll-bar frame. The cameras must be protected if the car crashes.

Mechatronics

Mechatronics is a term first used in Japan in the 1980s when robotic technology required engineers with multiple skills.

One of the engineering specialists at Globecam is Scott. He is a mechatronics engineer who has expertise in many areas, including:

- mechanical engineering
- electrical and electronic engineering
- computer science.

Scott designs new mechanical components by using computer–aided design (CAD) software. Then he hand-builds and tests the mechanical components for the remote-controlled camera systems. His wide range of skills and his calm, focused attitude make him a very good problem-solver and a valued member of the team.

Scott machines mechanical components.

Scott tests his new foot camera system in between the foot pedals of driver James Courtney's V8 Supercar.

11 V8 Supercars Trucks

Massive B-double trucks or shipping containers (for overseas events) are packed with V8 Supercars and all kinds of equipment, such as computers, tools and cables. Imagine the checklists itemising everything that the V8 Supercars teams need to ensure a successful event!

When you walk into the parking area of a V8 Supercars event, it is crowded with many massive B-doubles. Only people with special passes are authorised to go into the parking lot and the pits.

A convoy of B-doubles arrives at a V8 Supercars event.

B-DOUBLE TRUCKS

A B-double is a prime mover truck that tows two massive, long trailers. Each V8 Supercars racing team uses B-doubles to transport equipment and cars between the 16 events each year. Usually, two cars will fit in one trailer.

A prime mover that tows one trailer is called a semi-trailer.

V8 Supercars are double-stacked for transportation from Abu Dhabi to Australia.

The B-double trucks become information sites at the V8 Supercars racetracks.

12 A V8 Supercars Event – New Zealand

In April 2011, a V8 Supercars event was held in Hamilton, New Zealand.

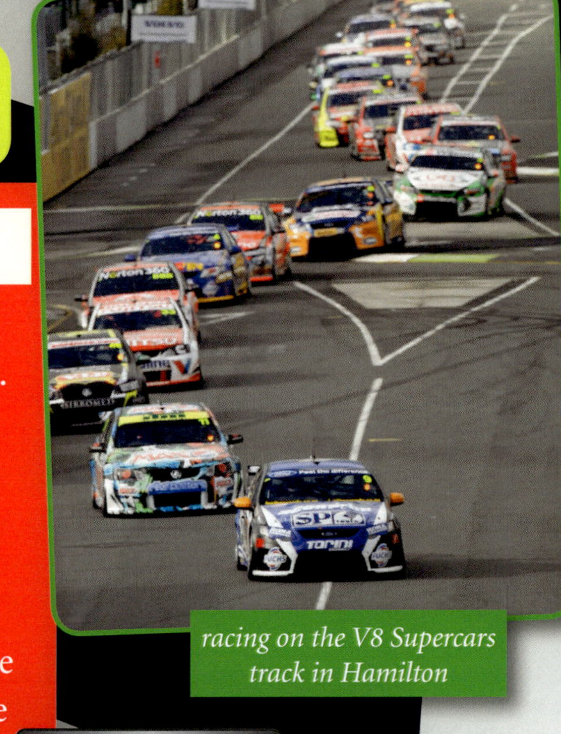

racing on the V8 Supercars track in Hamilton

V8 Supercar Shootouts

A Top Ten Shootout is contested at selected events as an extension of the qualifying races. The Top Ten Shootout involves the top ten qualifying cars racing in a one-lap dash. The Supercar with the fastest time assumes the pole position for the subsequent qualifying race. The remaining nine cars start the next qualifying race in their shootout places on the starting grid. The other 18 Supercars start the qualifying race in their positions – 11 to 28.

Racing Schedule

HAMILTON, NEW ZEALAND				APRIL 15–17
DATE	PRACTICE FORMAT	QUALIFYING FORMAT	RACE	RACE FORMAT
FRI 15 APR	40 x 30 minutes			
SAT 16 APR		1 x 20 minutes, shootout	5	1 x 200 km race
SUN 17 APR		1 x 20 minutes	6	1 x 200 km race

Racing Results

	1st	2nd	3rd
Saturday Top Ten Shootout	Jamie Whincup	Steven Johnson	Rick Kelly
Saturday Race	Rick Kelly	Craig Lowndes	Todd Kelly
Sunday Qualifying Race	Rick Kelly	Todd Kelly	Steve Owen
Sunday Race	Shane van Gisbergen	Lee Holdsworth	Garth Tander

New Zealand Driver Wins!

In April 2011, Shane van Gisbergen won his first V8 supercar race on the treacherously wet track. After skidding and crashing in the pit lane on the Saturday, his team managed to get his car ready for his milestone win the following day.

Shane van Gisbergen's V8 Supercar

winner, Shane van Gisbergen

Tander Skilfully Drives to Third!

Garth Tander is one of the most successful V8 Supercar drivers. At the 2011 Hamilton event, Garth Tander demonstrated his skilful driving as he navigated the 200-kilometre race from 19th position on the starting grid to finish in third place.

Garth Tander skilfully drives on the wet track.

Garth Tander's wife, Leanne Tander, watches a V8 Supercars race. Leanne is a successful V8 Supercars driver and in 2009 she competed against her husband in the Bathurst 1000 endurance race.

A massive television screen shows Garth Tander accepting his third-place trophy.

V8 Supercars Safety Messages

Safety is the number one priority for the V8 Supercars organisation and the teams. There are hundreds of ways that V8 Supercars address safety and here are just some of them.

Race Control Team

The race control team manages all of the race activities at each V8 Supercars event to maintain safety for everyone.

Safety Car

The safety car opens and closes each circuit, supplies race control with information about the track and leads the V8 Supercars field when track conditions require a controlled speed.

Thorough Car Maintenance

V8 Supercars teams dismantle, clean or replace many parts during and after each race.

Accurate Wheel Balance

The wheels on the V8 Supercars need to be properly balanced to maintain race safety.

First Aid

First aid organisations support other on-site medical teams, too.

Tyres Changed Often

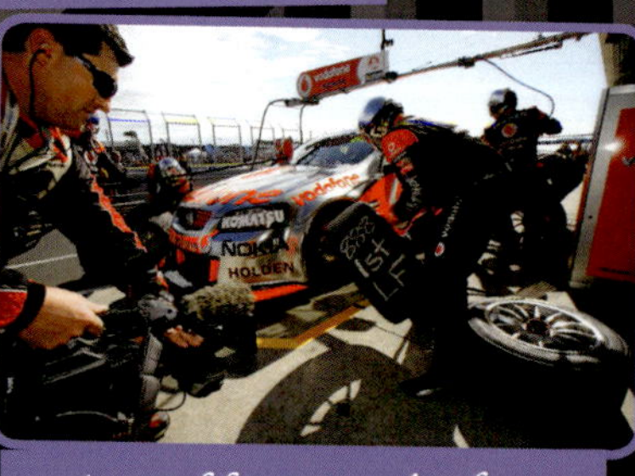

A set of four tyres is often replaced during the race in about 12 seconds!

V8 Supercars – Safety for Students

V8 Supercars Safety Messages

Roads are not racetracks.

V8 Supercars race under strict safety guidelines.

Students on Track

Schools receive free invitations to local V8 Supercars tracks to learn more about the V8 Supercars and their drivers.

Russell Ingall is the Road Safety Ambassador for V8 Supercars.

Russell Ingall, a successful V8 Supercars driver, visits a Queensland state school.

Junior Press Conference

Students who express an interest in interviewing and journalism can request an opportunity to interview V8 Supercar drivers about lifestyle, leadership, motivation, team building, and the importance of goal setting.

V8 Supercars drivers believe in teamwork and realistic goal setting.

V8 Supersafe

V8 Supersafe is a free driver safety program designed for students who are either eligible to get a driver's licence or who already have a licence to drive.

Primary school students can also attend the program to learn more about being a responsible passenger and road user (pedestrian or bicycle).

EXPOSITION FEATURE

More Motorsports Should Switch to Bioethanol Fuel

Motorsports around the world use massive amounts of petrol, which is a non-renewable fossil fuel that, when burned, causes harm to the environment. As a V8 Supercar mechanic, I believe that the best way to address this environmental problem is to reduce the use of petrol in motorsports. Bioethanol is the cleanest fuel for the environment, as it is made from feedstocks including sugar cane, potato and corn. A second alternative, which is better than straight petrol, is a blend of bioethanol and petrol.

1 One of the strongest reasons for restricting petrol use in motorsports is to do with greenhouse gases. Motorsports currently add significantly to the build-up of greenhouse gases in the atmosphere, which can contribute to global warming. By switching from petrol to bioethanol, emissions would be reduced to about half of those emitted by petrol-fueled racing cars.

sugar-cane stalks

sugar-cane field

Bioethanol Fuel

Jamie Whincup's head mechanic ensures that the right amount of bioethanol fuel is put into the tank during a pit stop. He measures the amount of fuel in seconds – not litres.

2

A second reason for using bioethanol fuel in motorsports relates to the preservation of Earth's non-renewable resources. Many people feel that using petrol for motorsports is an unnecessary waste of the world's fossil fuels. Bioethanol provides a much better option as it is a renewable resource that is produced from agricultural feedstocks. The crops use sunlight to grow and can be regrown after each harvest.

Refuelling

Jamie Whincup's team takes about 12 seconds to refuel his car and change over the tyres in the pits.

3

Finally, fuel made mostly from bioethanol is just as efficient as petrol and does not compromise race performance. This was proven in 2008, when the V8 Supercars organisation made a decision to supply only sugar-cane-based bioethanol to all V8 Supercars teams throughout their championship races. Performance testing of those cars showed that race results were not slower and that there was no damage to the engines. This proves that bioethanol is a viable alternative.

It is important that the world and its resources are looked after, not just for the current population, but also for future generations. Allowing motorsports enthusiasts to continue racing petrol-fueled cars is completely unnecessary, especially when there are environmentally friendly alternatives.

Index

Abu Dhabi 4, 20, 25

Australia 4, 7, 13, 14, 15, 20, 22, 25

B-double trucks 24–25

cameras 18–19, 20–21, 22–23

endurance race 15, 27

bioethanol fuel 7, 30–31

go-karts 10–11, 14

mechatronics 23

New Zealand 4, 6, 14, 16, 26, 27

Owen, Steve 8, 9, 14–15, 16–17, 18–19, 26

safety 6, 13, 16, 28–29

telemetry 7, 21

Top Ten Shootouts 26

V8 Supercars Championship 4–5, 8, 13, 14, 16, 22, 31

Whincup, Jamie 7, 8, 9, 10–11, 12–13, 15, 26, 30–31

Glossary

chassis The frame of a car, to which the body and wheels are attached

cylinders The part of a car's engine where the petrol and air is fired by a spark from a spark plug

dry ice The frozen or solid form of carbon dioxide

feedstocks Raw materials used in the manufacture of a product

pit lane An area at the side of a race-car track where cars stop for refuelling and repairs during a race

pit stop A brief stop made at a pit by a racing-car driver for refuelling and repairs during a race

pit stop tent A tent in which a racing-car crew waits for its driver to make a pit stop

pole position The position taken by a driver at the front of the starting grid on a race-car track

Other NLD Titles

ARTS NON-FICTION

A MUSIC DREAM COMES TRUE

FIREWORKS ART

TELEVISION TEACHES TOO!

DESIGNING FOR CINEMAS

A WRITER'S INVITATION INTO AN IMAGINARY WORLD

OUTSIDE BROADCASTS

ARTS BIG BOOK

ARCHITECTURE IS COOL

HUMANITIES NON-FICTION

THE LOGBOOK OF THE TIMANDRA

ECO-HEROES, AGENTS OF CHANGE

CAN YOU GO TO SCHOOL?

SMILE FOR THE CAMERA!

BMX RACERS

GOLD FEVER

HUMANITIES BIG BOOK

WHAT'S YOUR GPS?

NELSON
CENGAGE Learning™
For learning solutions, visit cengage.com.au

SCIENCES NON-FICTION

ANIMALS OF THE DARK
SOUND OUT THE DIGESTIVE SYSTEM
EXTRATERRESTRIAL (ET) SEARCH
SCIENCE OF FOOD TREATS
HIGH-VELOCITY V8S
WILD WEATHER REPORTING

SCIENCES FICTION

BARBECUE OVERLOAD
THE TREASURE SHIP OF ST ISABELLA
THE TALE OF THE TROUBLESOME TOAD
THE STOLEN FIGS OF GILGAL
THE GREAT GOLDEN NUGGET OF UTAH
THE RHYMACHINE OF RUFUS MCLEAN

SCIENCES BIG BOOK

NORTH POLE VERSUS SOUTH POLE

HIGH-VELOCITY V8S

Enter the exciting, action-packed world of V8 Supercars! As well as thrilling spectators with their high-speed racing, the V8 Supercars is an organisation that promotes road and passenger safety, and uses environmentally friendly bioethanol fuel. Find out how engineers use cutting-edge remote-controlled camera systems to bring detailed vision of the V8 Supercars action live to millions of viewers around the world every year.

Discover more about Nelson Literacy Directions at **www.cengage.com.au**

The Globecam team monitor their in-car camera systems at a V8 Supercars event.

For learning solutions, visit **cengage.com.au**

NLD Level: 12
Reading Age: 10.8–11.0
Text Type: Exposition

ISBN: 978-0170217484